Problems, Challenges and Investigations

by Tom O'Brien
Illustrated by Steve Gibbs

Claire Publications

Published by
Claire Publications
Tey Brook Craft Centre
Brook Road
Great Tey
Colchester
Essex CO6 1JE
Tel: 0206 212755

ISBN 1 871098 16 5

Printed in Great Britain

Contents

INTRODUCTION

The aim of this book is to encourage children to think.

Computers today can perform 12 billion operations per second, rendering certain algorithmic skills about as valuable as dinosaur-catching skills. Present-day pupils will spend most of their lives in the Twenty-First Century and their own children will be alive in the Twenty-Second. Thinking is the armoury they will need to use the machines to serve them best.

What does it mean "to think"? In general it means to construct ideas which maintain an equilibrium between one's self and the demands of the outside world. That's what this book is all about - engaging pupils in problem solving situations which call for the construction of novel and complex ideas. To bring the idea close to home, I ask the reader to look back on his or her own thinking. When were you last constructing ideas? What can you learn that might help children? It's a good bet that you've done some good thinking in problem solving situations, i.e., situations calling for novel and complex ideas.

TEACHING APPROACH:

The purpose of this book is to provoke pupils to construct important thoughts. The approach here is to pose problems for children to engage and thus to construct important mathematical relationships. Knowledge is thus viewed as provoked adaptation; that is, children construct ideas in the face of the challenge of a problem situation. The problems generally involve some sort of real-life context. This is a deliberate departure from the days when children learned facts and rules and procedures out of context and then -- if there were time at the end of the term -- they might have a few licks at applying them to some real world situations.

It is generally advised that pupils work together as collaborators -- in small groups of 2 or 3 or 4 -- in solving the problems posed in this book. It is generally advised that all of the tasks in this book are susceptible of hours, if not weeks, of thoughtful investigation, elaboration and extension. This is not the usual game of mathematical "exercises", i.e. exercise the little finger five times, then the ring finger, then the thumb, but never the whole hand.

But that's not at all to say that the teacher's role is one of passive observer. The teacher must be aware of the opportunities and purposes of the various problems and steer discussions into fruitful directions with appropriate gentle questions. A commentary for teachers, along with specific questions and suggestions for extensions, accompanies each of the problems in this book. It is NOT intended that the problems be one-shot ends in themselves. Rather, it is hoped that the problems given here will stimulate teachers' and pupils' generation of related problems and investigations and result in mathematical growth far beyond these printed pages.

ABOUT THE AUTHOR:

Tom O'Brien is a veteran teacher, teacher-of-teachers and curriculum developer with broad experience in research in pupils' mathematical thinking. He is also a parent with 75+ cumulative years experience who has widely and publicly credited his own children as "having taught him most of what he knows". He hopes that this book gives teachers -- "who are too often bossed around by idiots and who are generally, once unshackled, very very gifted" -- and pupils -- "who are widely regarded as inert vessels waiting to be filled, a totally unsupportable view" -- the chance to be agents, not patients, in the construction of important ideas. The Genevan psychologist Hermina de Zwart Sinclair said it very well says O'Brien: "We should all see children -- adults as well -- as wearing signboards which read UNDER CONSTRUCTION -- SELF-EMPLOYED".

ABOUT THE ARTIST:

Steve Gibbs is a graduate of Southern Illinois University. He works as a freelance artist/illustrator. He lives in Cottage Hills, Illinois and he enjoys bicycling, baseball and chess amongst other things.

It means to collect data and make sense of it.
It means to discern among "can't be", "might be", and "must be".
It means to make conjectures and hypotheses and test them honestly.
It means to represent -- in the sense of the Latin origin of the word (res praesantans), to make something present.
To think often involves abstraction, i.e., suppressing the particular details of this or that situation in order to consider many similar situations as one.
To think means to classify things, to order things, to hold several things in mind simultaneously and coordinate them and make judgements based on these coordinations.
To think, means to muster the intellectual resources -- the abilities and the skills and the attitudes -- to encounter and resolve novel situations and to invent new situations to engage.
It means to organize reality. From birth we ALL organize reality. One of the most important organizing jobs is done by children under age 5 all over the world who learn, without schooling, their own native language.
A British friend, David Fielker, said it very nicely, when we team-taught a class for teachers a long time ago: "We all find it easy to think of children as musicians. Shouldn't we also think of them as composers -- not mere players of someone else's music."
THINK

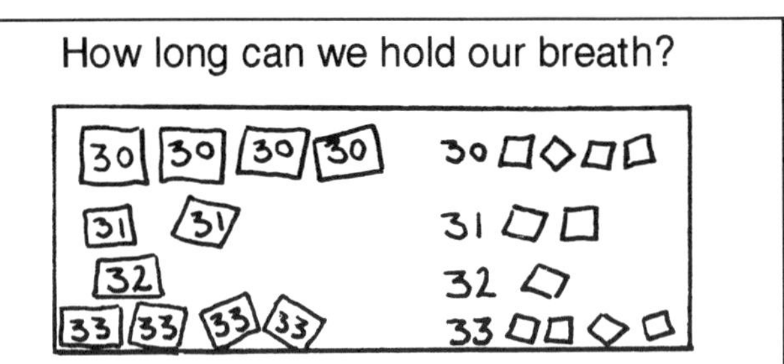

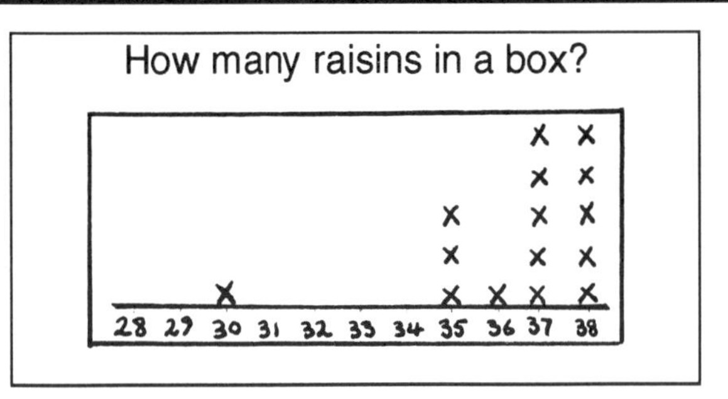

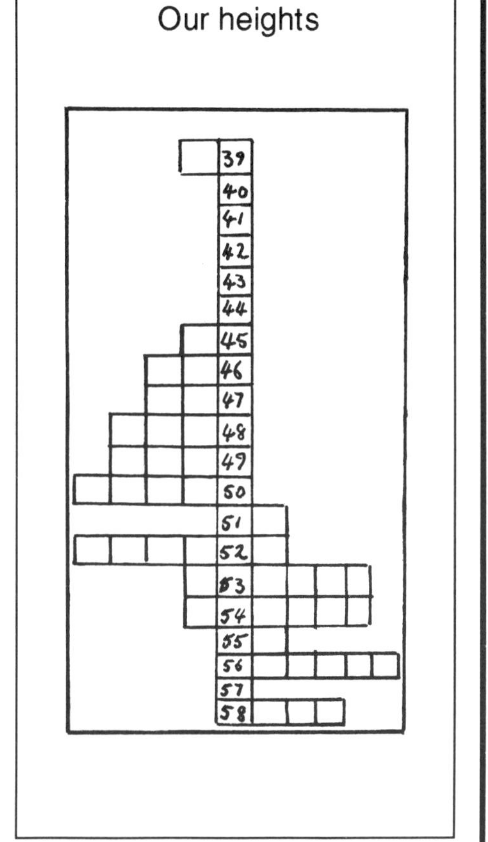

A NOTE ABOUT GRAPHIC REPRESENTATION

The Piagetian psychologist Hermina de Zwart Sinclair says that representation is a major intellectual act. The word comes from the Latin, res praesantans, to make something present -- visually, verbally, numerically, in poetry and music, even with mathematical things called ratios.

One of the major aspects of this book is that it involves a lot of data gathering, and it asks pupils to represent their findings -- to make their findings present. There are lots of interesting ways -- through a play, in a cartoon, in prose -- as well as the usual graphic approach, and the graphs need not be as dull and boring and flat as they have been historically.

Here are several different ways of "res praesenting" numerical data which may be new to the reader and to his or her pupils. These are drawn from "Used Numbers", a lovely set of booklets published by Dale Seymour Publications.

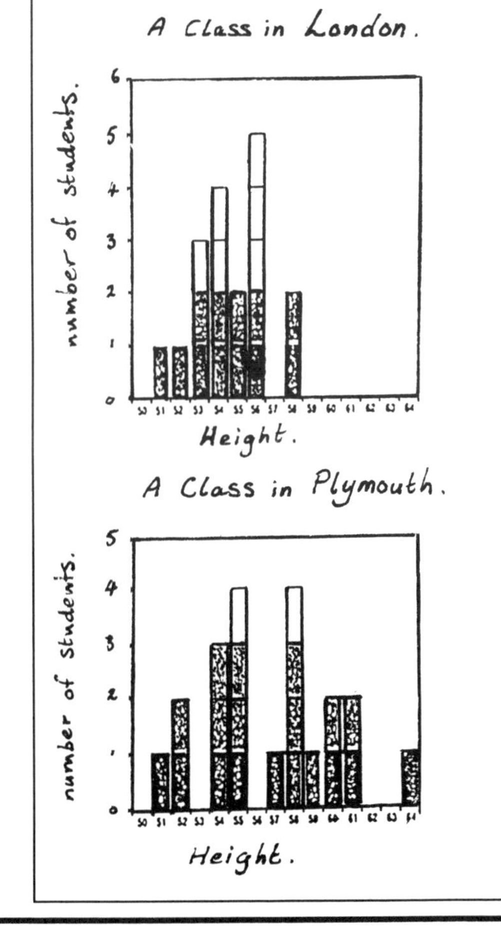

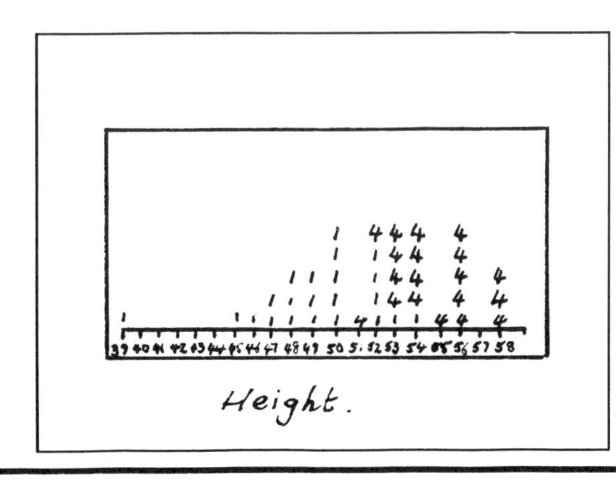

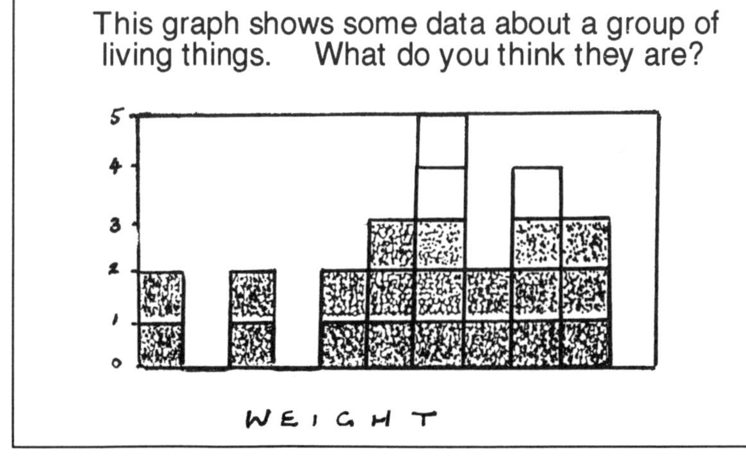

The highest recorded rate of public speaking occurred in a speech by President John F. Kennedy in December 1961: 327 words per minute.
(Source: Guinness Book of World Records)

What is the fastest you can speak articulately?
Test your classmates too.

HOW FAST DO YOU TALK?

PURPOSE

As is the case with a number of investigations in this book, children get a chance to gather data involving real-life issues. In this case, they are called upon to establish procedures for measuring words per minute. In addition, the nature of the investigation involves the notion of ratio. In the extensions, they are asked to make and test judgements regarding speed of speech and the context in which the speech was conducted.

COMMENTS

How to measure speed of speech? The issue is ratio, and thus the issue is a many-to-many or many-to-one correspondence. That is, one can set aside a one minute period and count how many words are said in that period. (But here is another problem. How to count words when they come out so fast? The use of a tape recorder may be helpful).

Possible follow-ups are abundant. Given this experience with measurement involving ratio, children might measure their walking or running speed (the ratio of feet or metres to a measure of time).

One interesting investigation conducted by a class of 10 year olds was the speed of cars passing their school and its relation to speed limit. They found that the local Police Chief exceeded the speed limit and so they had the opportunity of mixing maths with language by addressing a polite letter to the Police Chief at City Hall.

EXTENSIONS

1. What's the rate of speech of a television newscaster? A radio newscaster?
2. What's the rate of speech for you in normal conversation? How about your classmates?
3. What are the variations in rate of speech of a particular person (you choose the person) in various situations?
4. What relation is there between rate of speech and the situation/topic/audience? Make some guesstimates and test them out by gathering data.
5. What's the rate of speech for you and your classmates saying the following :
   a) Red leather, yellow leather (3 times)
   b) How much wood would a woodchuck chuck if a woodchuck could chuck wood? (3 times)

CLASSROOM QUESTIONS

1. What does it mean to say that President Kennedy spoke at a rate of 327 words per minute.
2. How can you work out someone's rate of speech?
3. What does "articulate" mean?
4. Is there variation among people?
5. What causes any variation that might occur?

Suppose three people were starting to play a game and they needed to choose which player would start.

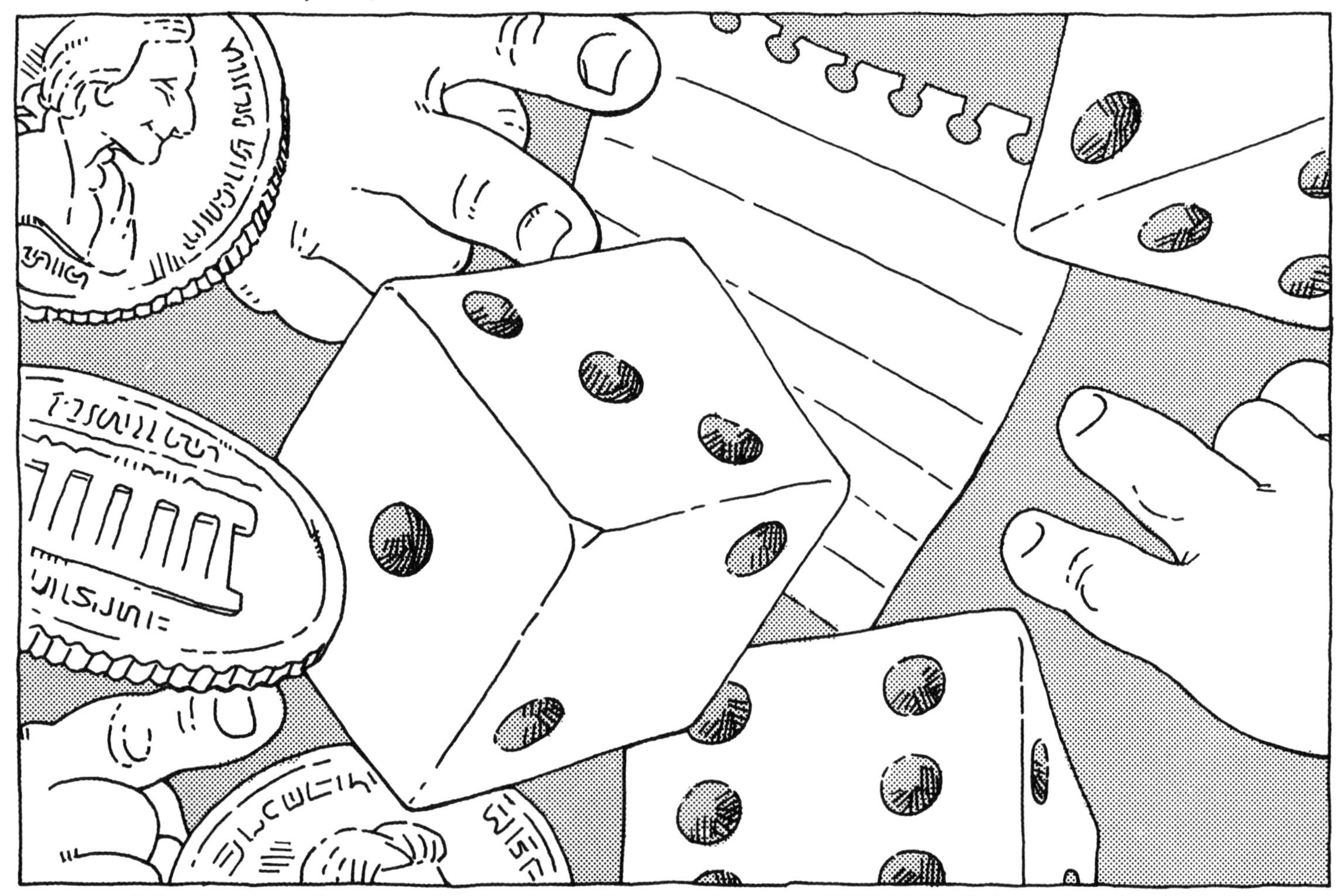

How could they choose a starting player FAIRLY using:
one die, two dice, three dice, one coin, two coins, three coins, their hands, a piece of paper?

PLAYING FAIR

PURPOSE

To devise rules so that dice and coin events are equally likely.

COMMENTS

This is not so easy as it looks. Three people having an equal chance with one toss of a die? That's OK -- Person A is assigned 1 and 2, Person B is assigned 3 and 4, and Person C is assigned 5 and 6. But what if you have two dice? The same 1-2, 3-4, 5-6 assignment? What about three people and two coins? Person A gets heads/heads, Person B gets tails/tails and Person C gets a mix?

Further, suppose that one can theoretically arrange for equally likely events -- as seems so easy with the 1-2, 3-4, 5-6 split-up for one die. Toss a die a thousand times (or simulate it with a one-line computer program) and measure the probability.

EXTENSIONS

A great many gambling games are played - dice, roulette, etc.

Are some of them more fair than others?

CLASSROOM QUESTIONS

1. What does being fair mean?

2. How do you know when something is fair?

3. What happens when you roll a die?
   What are the possible outcomes?
   Are the chances the same? Two dice?
   Three dice?

   How about the coin? Two coins? Three coins?

How many people in your town will hug someone today?

HOW MANY HUGS ?

PURPOSE

To involve pupils in :

a) data gathering in an everyday life situation

and then

b) to extend the local data to a bigger picture via proportional thinking

COMMENTS

This activity is one of many which involve data collecting in real-life contexts.

Some other investigations are :

1) How many auto accidents within 10 miles of your house today?
2) How many people in your city will eat ice cream today?
3) How many vegetarians in your town?
4) How many males? How many females on your street?
5) What proportion of people in your village are happy with their jobs?
6) What proportion of people in your village are over 65 years of age?
7) How many babies will be born in the world today?

EXTENSIONS AND ALTERNATIVES

1. Ask children what issue they would like to investigate.
2. How many hugs in your family in a day? Does one family differ from another? Why?
3. How many hands are shaken in a day?

There are 11 million sheep in Wales, four times the number of people.

In your village or town, where might you find ratios of 2 to 1? (Hands to people). Where else?

How about 3 to 1? 4 to 1? 5 to 1? 10 to 1? A hundred to 1? A million to one?

11 MILLION SHEEP IN WALES.

PURPOSE

To enable pupils to see fractions as many-to-many (or many-to-one) correspondences. At the heart of the fraction 1/4 is the notion "for every 1 there is 4" (in the problem involving sheep and Welshman: for every one person there are four sheep). Suppose the fraction were, instead, 3/4. The same notion would hold; for every three people, four sheep.

COMMENTS

This notion of "for every" or "for each" may go a long way toward reducing the mysteries that so becloud fractions. For example, the issue of equivalent fractions is simply a matter of many-to-many (or many-to-one) correspondence.

In the case of the sheep and the Welshmen, for example, it is easy to see that one for every four is the same correspondence as 2 people for every 8 sheep.

And, in general, a / b (which is a for every b) is the same as a times z for every b times z. (That is 1 for every 4 is the same correspondence as 2 for every 8, 3 for every 12, and in general 1 times z for every 4 times z.)

Now suppose you are asked in a "word problem" or a real-life investigation to figure out what 3/4 of 112 is. The question asks about 3 for every 4 in 112. There are 27 fours in 112 so the answer to the question is 27 threes, i.e., 81.

EXTENSIONS

1. In a certain magical village called Happywald, there are three Wimbles for every person. (A Wimble is a make-believe animal.) Draw a picture of Happywald.

2. What is the ratio of cars to people in your neighbourhood. Pets to people? Birthdays to people? Words to people (in an hour)? Eye blinks to people (in a minute)?

3. Happywald has 12 million cats which is four times its number of people. How many people are there in Happywald.

4. A class has 24 children, which is 1/4 (one fourth) the number of books in the library. How many books are there for each child? How many books are in the library?

5. A family has 36 chickens. This is 6 times the number of children in the family. How many children are in the family?

6. A school has 196 children. The children will go for a zoo visit. The transport vans hold 8 children each. How many vans are needed for the journey?

7. Take a look at Question 3 and make up some problems yourself.

Draw the car you think would appeal most to the following different kinds of people and list the features it must have.

A car for:

A teacher who doesn't like children;

A teacher who does like children;

An alligator trainer;

A forgetful person who always loses things;

A family of 20 who doesn't want to keep stopping for food and things;

Someone like you;

Someone like you who has won a competition and has unlimited funds for this car.

CRAZY CARS

PURPOSE

To mess about with coherence and incoherence in a visual medium.

COMMENTS

The morning that I (the author) wrote this page, I visited an eye doctor and had my eyes examined.

The doctor pointed to a certain eye chart and asked me to read the last line.

"Oh, Dear Heavens", I said. *

"No-ooo", said the doctor. "Try again".

The point of this is that while coherence is a major issue in one's life (lest one fly to pieces), moderate INcoherence is often quite appropriate. Moderate incoherence is often at the heart of humour ! And one cannot appreciate incoherence (and laugh) unless one has a sense of coherence in the first place.

Thus, this activity is intended to be fun, but it is also quite serious. Think about what life would be like if everything were coherent. What if everything were INcoherent? Ouch!

Children are likely to be shockingly, wonderfully, brilliantly inventive with this activity, especially the door-mat kids -- the children who are at academic risk because they are not memorisers.

This activity involves drawing, not print. All the more likely is it that the door-mat kids will shine through. Much of education, world-wide, comes, goes, and is measured in print.

A lot of people are not so good at print but they are very very good at visual imagery. My experience is that these children are vastly under-rated (if not ignored) in the world of ideas and expression.

(* That's not what I *really* said !)

The first successful gasoline-driven car was built by Karl-Friedrich Benz, and driven in Mannheim, Germany in 1885.

What is the most common make of vehicle in your community?

AUTOMOBILES

PURPOSE

To provide an arena for data-gathering.

COMMENTS

An investigation like this was undertaken by a friend of the author, a teacher with a class of very low-achieving youngsters. The pupils caught fire for the first time in their academic lives. Why? Because cars and car issues were important to them.

With the teacher's gentle guidance, the class ultimately extended their survey interests and their survey techniques to the school's lunch room. Their findings, presented politely to the lunch room staff, had a great effect in changing menu items and in general improving lunch room quality.

A potential follow-up to the original car investigation is to study the relation between make and age (and cost?) variables and demographic variables. For example, do rural people differ from city people in their car preferences? With a community, what difference does occupation make?

EXTENSIONS

1. What's the age of cars in your community? Do a survey and present the results.
2. Scavenger Hunt -- what's the most unusual car in your community? Do a search, then draw a picture of this unusual car.
3. Choose two parts of your town. Survey the cars in each. Are they different?
4. Design your own car. Draw a picture.
5. Design a car for someone in an unusual profession -- lion tamer, for example. Draw a picture.

CLASSROOM QUESTIONS

1. What is meant by "vehicle"?
2. What is the most common vehicle in your community?
   A car -- a truck -- a bike, etc.
3. Assuming the most common vehicle is a car, how could a survey be made to determine which is the most common make and model?
4. Can you think of another way to find out?
5. How many makes of cars are there on any one day in your town? On a given parking lot?

Capt. Wilson Kettle (1860-1963) is reported to have had 582 living descendants at the time of his death. (Source: Guinness Book of World Records)

Who has the largest number of living descendants in your family?

FAMILY GROWTH

PURPOSE

Aside from sociological considerations, this activity has a major mathematical payoff. The very nature of mathematics is the construction and investigation of relations. Family relations are a rich arena for children to begin (or extend) their ideas of relationships, ideas involving issues such as logical necessity, classification, and order.

This activity also provides an arena for real-life investigations.

COMMENTS

Here is a question for advanced pupils to engage: Every person now alive has two parents. And those parents each had two parents of their own. Given that this is so, it seems reasonable that there were many more people on earth 100 years ago (for example) than there are today. But population statistics don't support such a view. What's going on here?

EXTENSIONS

1. Who in your family has the largest number of living descendants? Make a chart to show your findings.
2. Who in your family -- cousins/aunts/uncles, etc. -- has the largest number of living relatives? Relatives by marriage do not count.
3. Who in your family (blood relatives only) has the largest number of living ancestors?
4. Pick an ancestor who is no longer living. Show the person's family tree.
5. Interview a neighbour. Ask him/her to choose one of his/her relatives who lived in 1900. Get enough information from him/her so that you can draw up a family tree for that person.

CLASSROOM QUESTIONS

1. Do you have any descendants?
2. Are you a living descendant?
3. How many living ancestors do you have? Make a chart to show them.
4. How many living relatives do you have? Diagram.

Pretend that you and your family are going camping from Friday 5:00pm to midday on Sunday.
Plan a shopping list for food.

CAMPING OUT

PURPOSE

The purpose of this activity is to provide an opportunity for children to conduct a real-life investigation involving cost-of-living. In order to do so, they need to plan ahead, to estimate, to compute, and to make judgements.

COMMENTS

This activity, as with most of the other activities and investigations in this book:

a) is not likely to be finished quickly,

b) is only a small sample of similar possible investigations which the teacher or the children might propose.

CLASSROOM QUESTIONS

1. What do we know about food needs on a camping trip? Quantity, type, etc.
2. Should we consider cost? Is it important that we know who will go along?
3. If you were leaving from home, would you need to buy everything you needed to take?
4. Would it help to make two lists -- things we want, and things we need ?

Who walked the greater distance?

WHO WALKED THE GREATER DISTANCE?

PURPOSE

This activity is concerned with children's keeping in mind and coordinating several different aspects of a situation.

COMMENTS

Again, this is a two-dimensional issue. It is not the case that the number of steps one walks determines one's distance covered. The size of the step is a variable which must be considered as well.

One way to view the investigation is that it concerns a move from absolutism to relativism. Younger children are likely to think that the number of steps is an absolute determiner of distance. The fact is that distance involves the relation between number of steps and size of step.

EXTENSIONS

1. Choose a particular spot on the school grounds (or in the school) for various members of the class. How many steps does it take to get there?

2. Do the same thing, but now keep track of time.

CLASSROOM QUESTIONS

1. What is your hunch about the answer -- justify your hunch?

2. Is it possible to know for sure which is farther?

3. Let's choose two of you. Go outside or into a corridor and see what the answer would be if you performed the task (deliberately choose two kids of different height/leg length).

4. What if these two took the same number of steps -- say 100 or 200?

5. Let's compare everyone's step length. Is "step" a good unit of measure?

A television report said that Detroit is the potato chip capital of the USA. Detroit citizens eat an average 70 bags of them each year, while the average American eats 45 bags.

How many bags of potato chips do you eat in a year?

POTATO CHIPS

PURPOSE

Children gather data in order to solve real-life problems. But it is not likely that they will keep track of the potato chips they eat for a year. Rather, they may gather data for a week and then extend their findings to a year's duration. Similarly, in extension 2, it is unlikely that they can sample their whole town. Rather it is more likely that they will sample a few people (their own family, neighbours, relatives) and extrapolate (extend reasonably) their findings.

COMMENTS

Most "word problems" amount to mere computational exercises. Rarely is it the case that children have to provide some information of their own in order to tackle a problem. In extension 1, the problem solver can only complete the task when an estimate of the cost of chips is known. One can obtain this data from local food advertisements, from a visit to the local food shop etc. But the problem is not quite that simple. Such an estimate, it may become clear, is probably highly dependent on the size of package one samples for price information.

And so this activity opens the way for another investigation -- the variability of prices of a given item, given various packaging. In food buying, is the larger package always a bargain? Is the "economy size" always economical? (Aside from the possibility that a larger package may result in left-overs and waste, it is sometimes the case that the "economy size" is more expensive per unit than other sizes!)

This activity opens the path for a number of other follow-ups. If Detroit is the potato chip capital of the world, what other capitals might there be? Beer? Soup? Lentils? Ice Cream?

In extension 3, how might one make an estimate of the data for an entire town? It would take years to survey large towns! One solution is to make an estimate from a small sample. A simple example is the following: Suppose a "town" consists of ten families. And suppose one samples two families and finds out that they consume 1 pound of crisps per year. It is reasonable if the sample is unbiased to estimate that a village of 10 families consumes five times as much as the two family sample.

EXTENSIONS

1. How much do you spend on potato chips each year?

2. Gather some data and figure out how many bags are eaten by your classmates.

3. What's the case in your town as far as chips are concerned?

4. What's the weekly expenditure in your family for meals? Can you break this overall meal expenditure into categories (such as snacks, etc.)?

CLASSROOM QUESTIONS

1. Are all bags of potato chips alike? How might they differ?

2. If all bags contained equal weights of chips, how many does the average Detroiter use per week? Per day? How about the average American/European?

3. How many chips are in a bag? How many chips are in a serving? Are there differences in brands?

How much does your family spend on food in a week?

HOW MUCH FOOD?

PURPOSE

As with several other activities in this book, this investigation helps children to see themselves as part of a human and contextual fabric, not as an atom unconnected with anything else in the world. The task involves real-life data gathering as well as estimation, classification and computation.

COMMENTS

Here are some categories of family expenses. Make an estimate of the weekly expenses in each category. Then gather data to see how close an estimate you made.

1. Food
   - i cereals, bakery products
   - ii meats, poultry, fish, eggs
   - iii dairy products
   - iv fruits, vegetables
   - v sugars, sweets
   - vi fats, oils
   - vii other
2. Housing
   - i rent
   - ii mortgage payments
   - iii fuel, other utilities
   - iv electricity
   - v household furnishings
3. Apparel
4. Transportation
   - 4.1 Private
     - i automobiles
     - ii fuel
     - iii auto insurance
   - 4.2 Public
5. Medical Care
6. Taxes

EXTENSIONS

1. The activity asks pupils to gather data on family living expenses. The data would make a rich source for a database. Here are some questions to ask:
   How does family size influence the family expenses?
   How does the age of children influence a family's expenses?
   Any differences between home owners and renters?
   Children may suggest interests of their own to investigate.
2. As with all of the data-gathering activities in this book, the suggested investigation can be a jumping-off spot for other explorations involving one's family "spending". How about time spent watching television? Time spent sleeping? Money spent on "needs" versus "wants". What's the difference?
3. The elaborate outline of possible categories (in the suggested extension) can be abridged as appropriate. One category scheme is simple: 1) Food, 2) Housing, 3) Clothing, 4) Transportation, and 5) Taxes. Pupils may wish to invent and use their own category scheme.
4. It may be fruitful, once their data-gathering is complete and their results organized, to have children compare their original estimates with the actual findings.

CLASSROOM QUESTIONS

1. Without asking the person who does the shopping for your family, how could you begin? It would be interesting to estimate before beginning and to revise the estimates several times during the project.
2. Is every week the same?
3. What makes some weeks more expensive than others?
4. What is the best sort of week to choose for this problem?

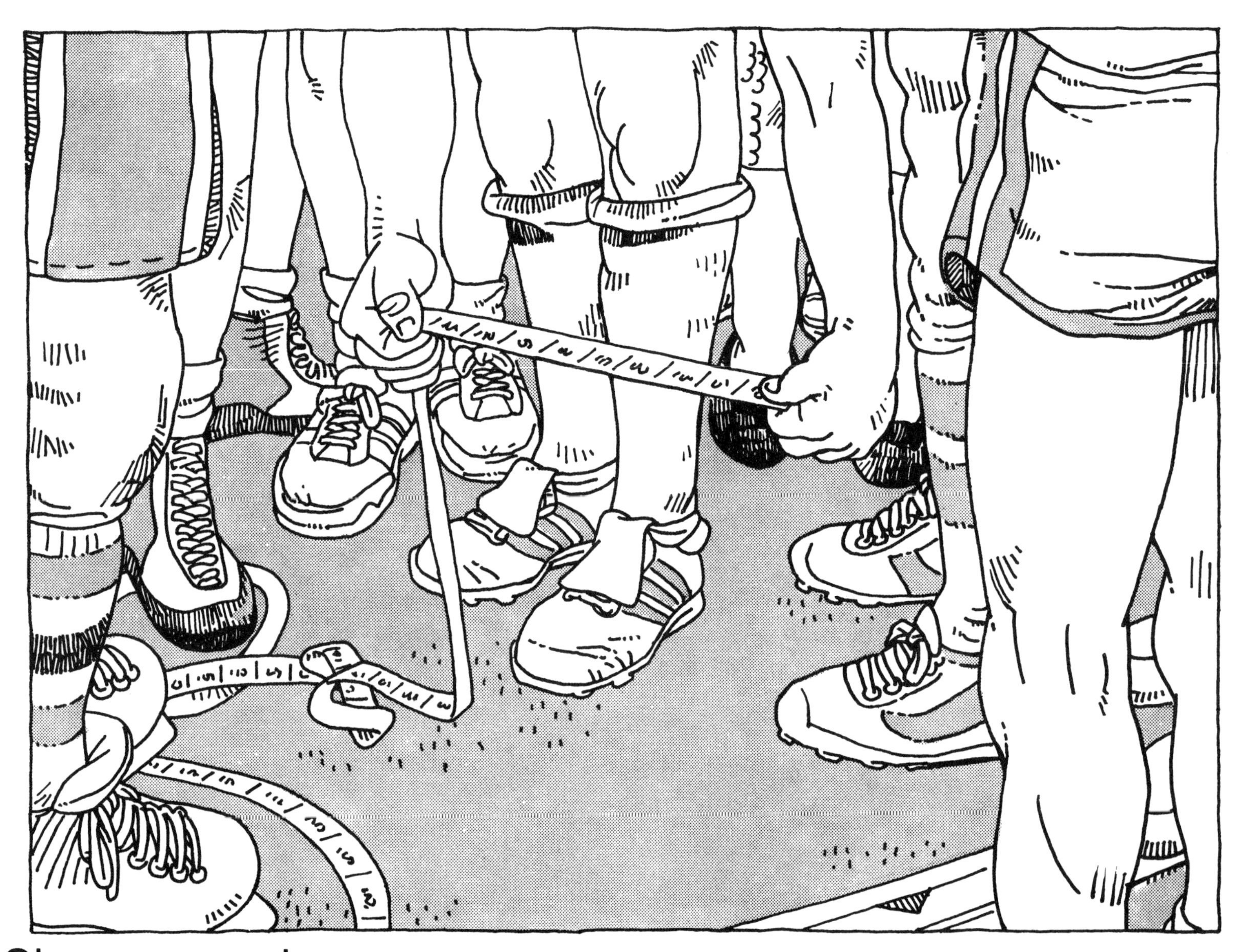

Choose a sport.
In an ordinary professional game, how far does a player run?

RUNNING

PURPOSE

This activity gives children the opportunity to conduct (and in an extension, to invent) an investigation involving data gathering, to make an estimate, and to test that estimate (and refine it if need be) in the face of data he or she has gathered.

COMMENTS

As with other activities, children have to devise a method of investigation. How does one measure distance? If the event is a 100 yard dash, the task is trivial. But what about football? (Might counting a player's steps help give an estimate of distance?)

Teachers may want to take care that the investigation children devise is a good fit for their abilities -- not too easy, not too difficult.

A fruitful and practical follow-up may be a conversation with a sports medicine expert. For the present pupils to stay active athletically throughout life, what precautions might they take? What sports activities can a pupil begin now which will give the greatest chance of sustaining him or her throughout life?

EXTENSIONS

1. How far do you run (or walk) in a day?
2. Make up some "distance" investigations of your own.

CLASSROOM QUESTIONS

1. Which sports might we choose? Teachers may choose to discuss with the students what advantage/disadvantage each sport has with respect to the task.
2. Will we have to play or watch a whole game to work it out?
3. Do all sports involve running?
4. Do all sports involve winning?

Smith is quite a common name. What does it mean?

What does your surname mean?

WHERE DO NAMES COME FROM?

PURPOSE

There is a rich history, a rich fabric of language and events and persons and places to which today's children are heirs (and in which they are participants).

This investigation immerses children in events which preceded them by centuries, events which may have considerable influence over their present life. It involves history (including historical accident -- the Duke of Such-and-Such invaded the Earl of Someplace Else with consequences which endure to this day), causation, classification, and where historical facts are unavailable plausible inference.

COMMENTS

The New Yorker magazine, September 10 and 17, 1990, contains an extraordinary portrayal of the drug life in New Haven, Connecticut. Terry, a 15-year old drug dealer "went Yale". He wore Yale sweat pants, a Yale wind-breaker, and a Yale basketball cap. The author, William Finnegan, conveys the central fact of young Terry's life: He has virtually no connection with the state of the world before he was born. To Terry, the world "Yale" is a designer label, like Armani or Gucci or Reebok.

The New York Times (September 26, 1990), reporting on the New Yorker article, says of Terry, "His charm and grace and lack of guile are striking. There is a child in there somewhere, beneath the drug dealer, but he has had to invent himself in the only world he knows. The finished product is by turns engaging, terrifying, heart-breaking."

Cannot we educators help children know (and extend) the civilized world rather than their inventing a world of their own <u>de novo</u>?

EXTENSIONS

1. The name Smith is very common. What other names are common? What do they mean?
2. What is the most common surname in your school? What does it mean?
3. What's the most common name in your town? What does it mean?
4. Varga is a very common name in Hungary and it means shoemaker . Choose other countries and find out what their most common names mean.

CLASSROOM QUESTIONS

1. What do you think would be the most common name in India, China, New York? How could you find out?
2. Why do some people have only one name -- or a hyphenated name? How about people in other countries?
3. How did people get their names originally? Does this help know meanings?

If the pattern continues black, white, black, white: are these squares black or white? 14th? 26th? 41st? 83rd? 144th? 1007th?

WHAT COLOUR IS IT?

PURPOSE

This activity asks children to perceive regularity and to use principles of regularity (odd/even), for example, to answer the questions rather than to employ brute-force counting.

COMMENTS

Research by the author shows that this is a very difficult problem for children under 12 so far as their knowledge of patterns is concerned. (Fair enough, many of these kids had had a mathematical diet of straight computation.) What many of them do -- even for the odd-even pattern -- is to count out the numbers one by one rather than to say something like "Oh, obvious, 144 is even and therefore . . . " This is not at all to say that these children are wrong. It is to say that their ideas are under construction.

The first pattern in the extensions is a simple times table. The second is a simple times table, but it starts out of phase. The last pattern is a first step toward a Fibonacci pattern. (A Fibonacci pattern goes 1, 1, 2, 3, 5, 8, ..., such that any term is the sum of the last two terms.) For further work with Fibonacci, turn your children loose on "Fascinated by Fibonaccis" by Trudi W. Garland from Dale Seymour Publications.

EXTENSIONS

Answer the same questions for these patterns.

1. 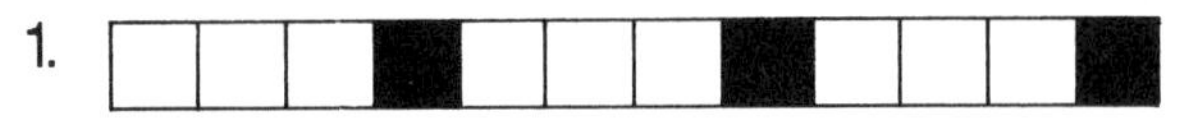

2. 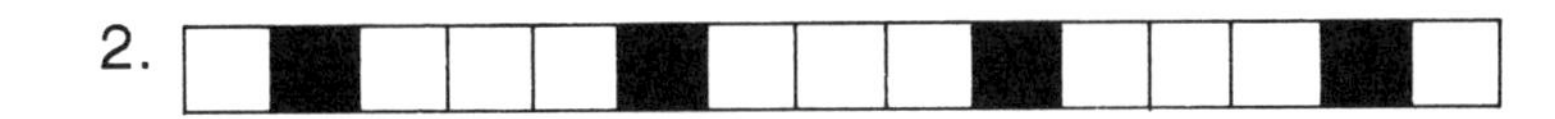

3. 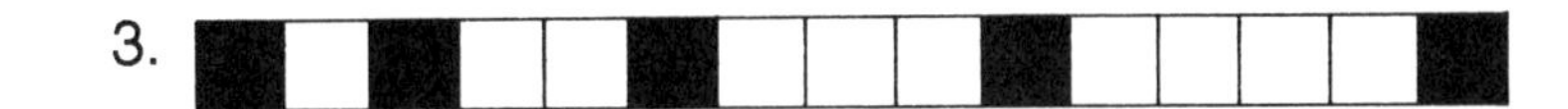

4. Make up some patterns yourself. Challenge classmates to determine what 1000 and 1 million might be.
5. Make up a pattern where 1000 is black.
6. Make up a pattern where 10 is black, 100 white, 1000 black and so forth.

CLASSROOM QUESTIONS

1. What's your guess?
2. How can you prove if your guess is correct?
3. Are there any other ways?
4. What about a different pattern?

Form teams of four.
Locate (but not necessarily bring to your classroom) the following items.

1. five foreign magazines.
2. a Citroen.
3. a foreign newspaper.
4. a family photograph of when your grandparents were young.
5. five interesting number plates.
6. a unicycle.
7. two French coins.
8. four menus from ethnic restaurants.
9. two newspapers at least 20 years old.

The first team to locate all the items is the winner.

SCAVENGER HUNT

PURPOSE

Present day children will live most of their lives in the Twenty First Century. (Their children will be alive in the Twenty Second Century). One of the necessary conditions for success and enjoyment in a fast moving and rapidly changing culture, unless one is a sea slug, is the ability to meet and resolve novel situations, especially problem situations. This scavenger hunt, as is the case with several others of the sort in this book, calls for children to muster their intellectual armament to engage and resolve new and challenging problems.

The purpose of these scavenger hunts is to give children opportunities to confront the unknown or the little-known. In doing so, children construct and extend their inference and classification abilities, not to say anything of flexibility, patience, knowledge of relationships, and general culture.

COMMENTS

Obviously children can suggest interesting items themselves. How about two teams, each of which makes up a list of interesting items for the other to find?

I once had the great fortune to ask Professor George Polya, one of the major thinkers and writers on problem solving, "What is a good problem?"

I paraphrase : "A good problem should first of all be difficult enough, else it is not a problem. Difficulty is not an absolute thing. It lies in the tension between the problem-solver and the demands of the problem."

"Second, the problem should be interesting. Once again, interest is not an absolute. What is interesting for one person may be terribly boring for another person.

"Third, and most important, a problem should go somewhere! That is, the problem shouldn't be merely an entertaining puzzle. It should engage the problem solver in issues which are important mathematically." (May I add culturally, intellectually, practically?)

Thus, if children do make up their own scavenger lists, it is fruitful for the teacher to scan the lists to see that they are not merely difficult (or obscure) but that they challenge the pupil to "go somewhere".

Teachers may be interested in reading Polya's classic book How to Solve It, which has been translated into more than 20 languages and which should be available in one's local library. It requires no more than a decent secondary school background in mathematics and it is widely regarded as the seminal book on problem solving.

EXTENSIONS

1. Pupils make scavenger hunt lists of their own. The teacher edits the lists so that they "go somewhere".
2. A simpler list may be best to start with :

   a phone book; a set of keys; a bank note; a postage stamp; a map or broom.

CLASSROOM QUESTIONS

1. What does locate mean? Children may suggest that items have to be seen before they can be counted as 'located'. Given the demands of school situations, it may be sufficient for participants to tell where the items are, rather than bring them to the class.

Could your grandfather have danced with Queen Victoria?

COULD YOUR GREAT-GRANDFATHER HAVE DANCED WITH QUEEN VICTORIA?

PURPOSE

Aside from sociological considerations, this activity has a major mathematical payoff. The very nature of mathematics is the construction and investigation of relations. Family relations are a rich arena for children to begin (or extend) their ideas of relationships, ideas involving issues such as logical necessity, classification and order.

This activity also provides an arena for real-life investigation.

COMMENTS

Here is a possible map of the situation:

199?	1950	1920	1890
Child now	Father's birth	Grandfather's birth	Great-Grandfather's birth

It isn't likely that a pupil's grandfather would have been alive in the late 1800's. OK, the problem is solved.

But a more interesting problem may be the following: Is it possible for a living person to talk with someone who talked with someone who talked with someone . . . . . who talked with someone who danced with Queen Victoria? (For example, right now, I can talk with my neighbour, who is 90+ years old. Let's assume that that person would remember conversations back to 1910, when she was a teenager. And let's assume that she talked (in 1910) with someone who was then 90 years old. That person may have been able to remember conversations dating back to 1830!

EXTENSIONS

1. What's the oldest memory in your family? (Do interviews of senior family members.)
2. Invent a family member, Susan, who lived 1900-1957. Interview family members to gather information regarding the important world events she lived through. Draw a biography of Susan's life.

CLASSROOM QUESTIONS

1. Who will you talk to first ?
2. How will you keep a record of what you find out?
3. How will you show what you have found out, in writing, on a chart as a cartoon?

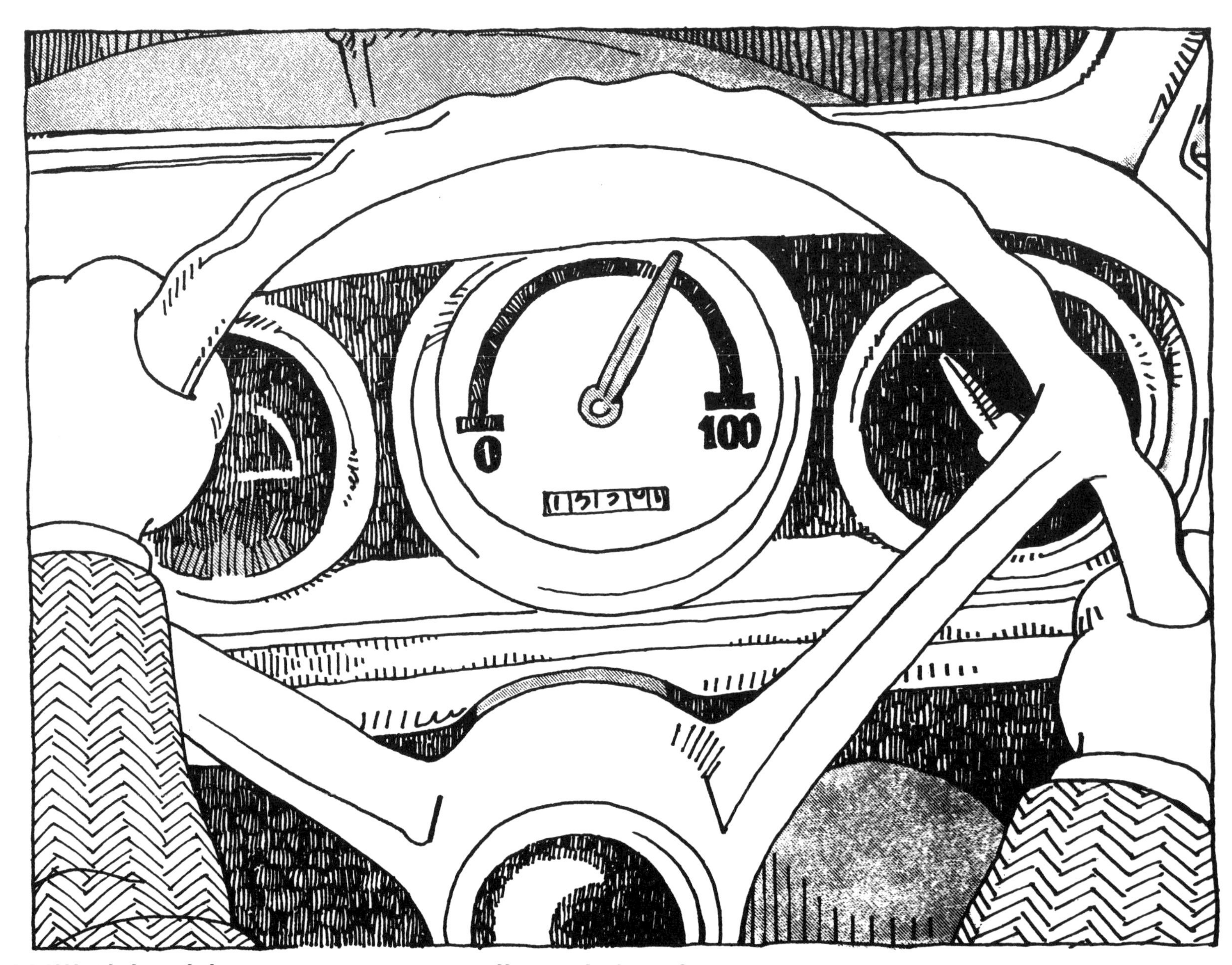

Will this driver get a speeding ticket?

HOW FAST IS THIS CAR DRIVING ?

PURPOSE

Here is an environmental investigation involving measurement of an important real-life variable, speed.

COMMENTS

It's often the case that one needs to infer information, given an incomplete picture. That's the case with the principal activity here, where pupils are asked to devise a method of estimating a speedometer reading. The fact is that this is a simple ratio situation. The speed readings range from 0 to 100 in just the same way that the central angle (the angle made by the needle and the horizontal) ranges between 0 and 180 degrees. That is, a needle making a 90 degree angle would point to a speed of 50. An angle of 45 would yield a speed of 25 (45 is 1/4 of 180, so 1/4 of 100 is 25).

A question of children: the scale of the speedometer is not given. Is it in miles per hour or km/hour? There's no way of telling but by context. It would be a pretty unusual automobile whose top speed was 100 km/h.

EXTENSIONS

In the extensions, children are asked to derive a measure of speed using their own techniques. This is fundamentally different from merely reading (or estimating) the speed shown by a speedometer. This issue of a two dimensional variable, speed, which is the ratio of distance to time. That is, for every so and so many units of time, one covers such and such units of distance.

1. What is <u>your</u> speed when walking?

2. What is your speed when running as fast as you can?

3. What is the fastest speed of various animals -- an ant, a turtle, a dog, a cat, a gazelle, an elephant?

CLASSROOM QUESTIONS

1. Do you think this driver would get a speeding ticket?

2. What is a speed limit? Why do they exist ?

3. What speed is shown?

4. Suppose the scale were in km/hr?
   What of miles/hour?
   How do you convert from one to the other ?

COST OF LIVING IN SELECTED CITIES

New York City = 100

Kinshasa	97
London	111
Montreal	87
Tokyo	154
Warsaw	41
Bonn	116
Cairo	84
Copenhagen	111
Geneva	137
Kingston, Jamaica	68

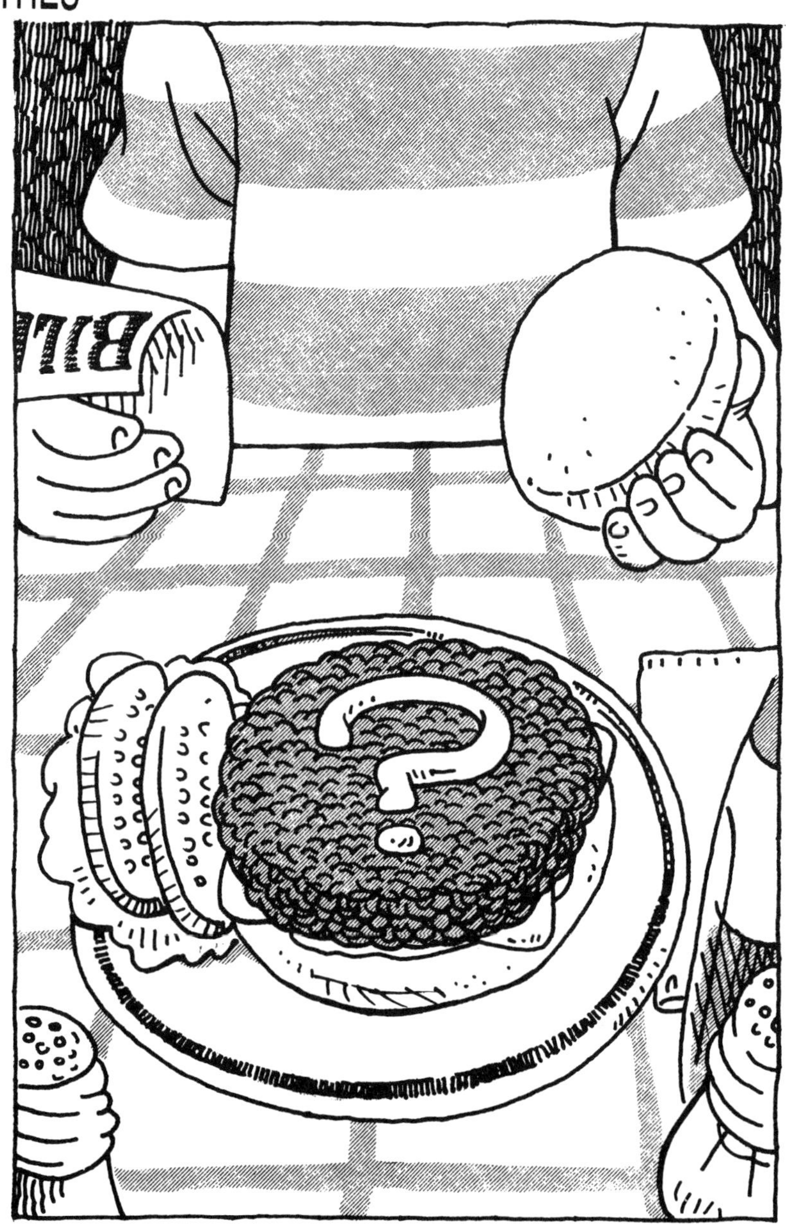

How much would a hamburger cost in each of these cities ?

LIFE IN THE BIG CITY

PURPOSE

To give children experience outside their own surroundings.

To provide real-life problems involving ratio.

COMMENTS

One powerful tactic in problem-solving is "If the original problem is overwhelming, try to solve a related simpler problem".

In starting to solve the original problem, it might be easier to solve a different problem: Suppose a hamburger cost £1.11? What would the Geneva equivalent be? What if it cost $1.00? Some clarification of the ratio-type table may be necessary. What the entries mean is that an item costing 100 units of currency in New York costs 137 of the same units in Geneva (and so forth for the other cities).

EXTENSIONS

1. What's the range in price of a hamburger in your town?
2. How about other foods -- chips, for example. What's the range in price?
3. Choose a particular item -- for example, Cola drink. What's the range in price in supermarkets?
4. Choose a city on the list of cities and make up a budget for a week's holiday in that city.

CLASSROOM QUESTIONS

1. What do the numbers in the chart mean?
2. What additional information will you need to answer the questions? How will you get this information?
3. Where is each of these places? Can you find out if a hamburger would be easy to buy there?
4. What do you think determines the cost of an item? Are the reasons always the same?

The fertility rate of various countries, taking into account births and deaths is as follows:

COUNTRY	1990 POPULATION (in millions)	POPULATION will double in this many years
Brazil	150.4	36
China	1119.9	49
France	56.4	157
Japan	123.6	175
India	853.4	33
Iraq	18.8	18
Sweden	8.5	311
USA	251.4	92

Make up some interesting questions about this data.

PEOPLE OF THE WORLD

PURPOSE

To develop ways to answer questions dealing with statistical data.

To use statistical data to construct and test hypotheses.

COMMENTS

As with all problems in this book, the use of calculators is by all means to be encouraged.

But be cautious that children do not lose sight of the magnitude of the numbers (reporting an answer as 300, for example, when it should be 300 million).

EXTENSIONS

1. Children may wish to explore the issue further by obtaining data on other countries or on other groups of people -- their own town, for example.

2. A question for further exploration: Why do the growth rates differ from country to country? One possibility is that bigger countries double their population in a larger number of years than smaller countries. Does the rest of the data support such an explanation? What factors influence population growth rate?

3. Some arithmetic exercises. Encourage estimates.
   - i. China has how many more people than India?
   - ii. The USA has how many more people than Brazil?
   - iii. The USA population is how many times that of France?
   - iv. The French population is how many times that of China?
   - v. In what year will the population of the USA be twice that of 1990?
   - vi. In what year will the population of Sweden be doubled?
   - vii. Assuming the same growth rate, when will Brazil's population equal that of the USA?

CLASSROOM QUESTIONS

1. What does the chart show?
2. What is a fertility rate?
3. How would this information be gathered?
4. Is there a country listed where population <u>isn't</u> increasing. Would this be possible?
5. How could we find out the fertility rate for our town? Our country?
6. Who would be interested in data like these? How might they be used?

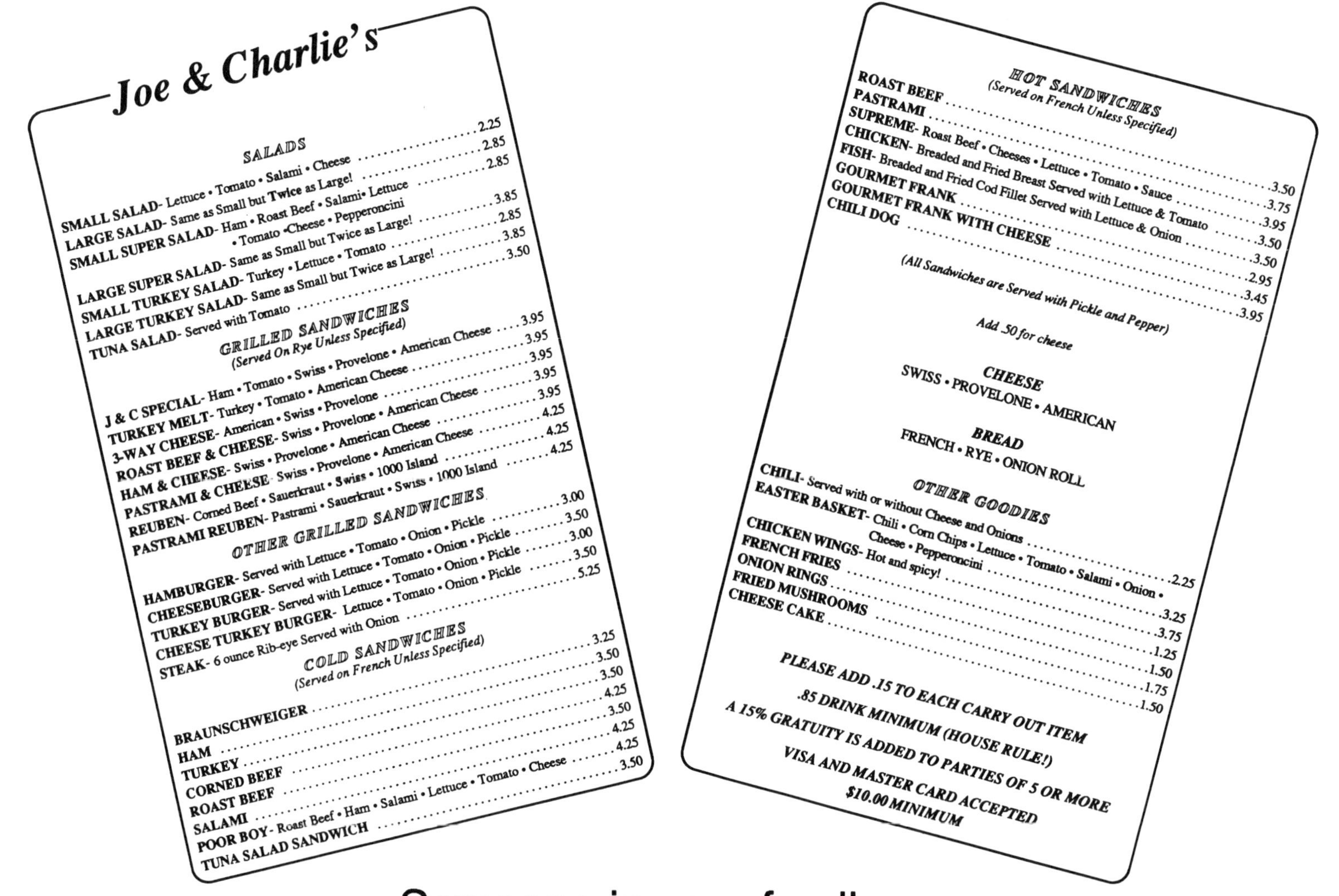

Joe & Charlie's

SALADS

SMALL SALAD- Lettuce • Tomato • Salami • Cheese 2.25
LARGE SALAD- Same as Small but **Twice** as Large! 2.85
SMALL SUPER SALAD- Ham • Roast Beef • Salami• Lettuce • Tomato •Cheese • Pepperoncini 2.85
LARGE SUPER SALAD- Same as Small but Twice as Large! 3.85
SMALL TURKEY SALAD- Turkey • Lettuce • Tomato 2.85
LARGE TURKEY SALAD- Same as Small but Twice as Large! 3.85
TUNA SALAD- Served with Tomato 3.50

GRILLED SANDWICHES
(Served On Rye Unless Specified)

J & C SPECIAL- Ham • Tomato • Swiss • Provelone • American Cheese 3.95
TURKEY MELT- Turkey • Tomato • American Cheese 3.95
3-WAY CHEESE- American • Swiss • Provelone 3.95
ROAST BEEF & CHEESE- Swiss • Provelone • American Cheese 3.95
HAM & CHEESE- Swiss • Provelone • American Cheese 3.95
PASTRAMI & CHEESE- Swiss • Provelone • American Cheese 4.25
REUBEN- Corned Beef • Sauerkraut • Swiss • 1000 Island 4.25
PASTRAMI REUBEN- Pastrami • Sauerkraut • Swiss • 1000 Island 4.25

OTHER GRILLED SANDWICHES

HAMBURGER- Served with Lettuce • Tomato • Onion • Pickle 3.00
CHEESEBURGER- Served with Lettuce • Tomato • Onion • Pickle 3.50
TURKEY BURGER- Served with Lettuce • Tomato • Onion • Pickle 3.00
CHEESE TURKEY BURGER- Lettuce • Tomato • Onion • Pickle 3.50
STEAK- 6 ounce Rib-eye Served with Onion 5.25

COLD SANDWICHES
(Served on French Unless Specified)

BRAUNSCHWEIGER 3.25
HAM 3.50
TURKEY 3.50
CORNED BEEF 4.25
ROAST BEEF 3.50
SALAMI 4.25
POOR BOY- Roast Beef • Ham • Salami • Lettuce • Tomato • Cheese 4.25
TUNA SALAD SANDWICH 3.50

HOT SANDWICHES
(Served on French Unless Specified)

ROAST BEEF 3.50
PASTRAMI 3.75
SUPREME- Roast Beef • Cheeses • Lettuce • Tomato • Sauce 3.95
CHICKEN- Breaded and Fried Breast Served with Lettuce & Tomato 3.50
FISH- Breaded and Fried Cod Fillet Served with Lettuce & Onion 3.50
GOURMET FRANK 2.95
GOURMET FRANK WITH CHEESE 3.45
CHILI DOG 3.95

(All Sandwiches are Served with Pickle and Pepper)

Add .50 for cheese

CHEESE
SWISS • PROVELONE • AMERICAN

BREAD
FRENCH • RYE • ONION ROLL

OTHER GOODIES

CHILI- Served with or without Cheese and Onions 2.25
EASTER BASKET- Chili • Corn Chips • Lettuce • Tomato • Salami • Onion • Cheese • Pepperoncini 3.25
CHICKEN WINGS- Hot and spicy! 3.75
FRENCH FRIES 1.25
ONION RINGS 1.50
FRIED MUSHROOMS 1.75
CHEESE CAKE 1.50

PLEASE ADD .15 TO EACH CARRY OUT ITEM
.85 DRINK MINIMUM (HOUSE RULE!)
A 15% GRATUITY IS ADDED TO PARTIES OF 5 OR MORE
VISA AND MASTER CARD ACCEPTED
$10.00 MINIMUM

Someone in your family
is celebrating a birthday.
You decide to eat out.
How much will it cost?

HAPPY BIRTHDAY

PURPOSE

To give children real-life experience in estimating.

COMMENTS

Teachers will probably wish to invite the children to bring in some menus to plan the birthday party.

EXTENSIONS

1. Estimate the total cost of a restaurant outing for your family. Then gather data and check your estimate.
2. Estimate the cost of a wedding, with all accompanying events, parties, dinners. Then gather data. Are you a good estimator?
3. In Questions 1 and 2 who in your class made the best estimate? What tactics did the good estimator(s) use?

CLASSROOM QUESTIONS

1. What extra information do you need to solve the problem?
2. Are there likely to be any costs other than those on the menu?

Draw a cartoon biography or a family tree for the following newspaper story.

Dubinsky, Ida (Lipsitz). Wife of the late Jack. Beloved mother of Phyllis and Horace Levin. Adored grandmother of Karen and Andy Abrams, Stu and Ellen Levin. Great-grandmother of Lorie and Jason Abrams and Alison Levin. Funeral private. Shiva in the Abrams home.

Draw several possible family trees.

FAMILY TREES

PURPOSE

Today's world is such that many children see themselves as cut off from others, as atoms unconnected with any molecule. This activity helps children to see themselves as part of a fabric. Children need to know the importance of relationships -- the world doesn't consist of unrelated atoms -- and people relationships and especially family relationships are both psychologically and intellectually important.

Aside from such sociological considerations, this activity has a major mathematical payoff. The very nature of mathematics is the construction and investigation of relations. Family relations are a rich arena for children to begin (or extend) their ideas of relationships, ideas involving issues such as logical necessity, classification, and order.

This activity also provides an arena for real-life investigation.

COMMENTS

One very simple definition of thinking is "generating alternatives". Sometimes we teachers tend to foreclose alternatives in the interest of efficient education. This is especially true in mathematics teaching, where it sometimes still is believed that there is one and only one solution to a given problem.

This activity calls for children to generate possible scenarios for the complex relationships described in an obituary. (The obituary was taken from a local newspaper. The names were changed for privacy's sake).

This problem is quite difficult. How can Ida Lipsitz Dubinsky be the mother of Phyllis and Horace Levin?

EXTENSIONS

1. Pupils may profit from tackling other obituaries and from creating obituaries for other children to explain.
2. What is a shiva? What commonality, if any, does a shiva have with other rituals or customs?
3. What do Hindus do? Spaniards? French? American Indians? Mexicans? Irish?

CLASSROOM QUESTIONS

1. Have you ever read anything like this?
   What is it called?
   Where might you find one?
2. Have you ever seen a cartoon biography or family tree?

A newspaper article described the change in the price of bread from 0.8 cents (US) to 2 cents (US) in Egypt. Hamid, an Egyptian who lives in Boulaq, said that this news was devastating. Each member of his family eats a loaf of bread each day and Hamid's monthly income is $10 (US) per month.

Do you think this news was devastating?

LOAVES OF BREAD

PURPOSE

Children are presented with a real-life situation quite foreign to their own and are asked to put the situation in context.

COMMENTS

This is a true story taken from a newspaper in July 1990. Of course Hamid was upset. The monthly cost of the newly priced bread for his family is now more than 30% of his income!

A cultural extension of this activity -- and perhaps an eye-opener for many children -- might be to investigate and compare the data on income for various countries of the world (or for various parts of their own country).

EXTENSIONS

1. How do pupils in the class spend their money?
2. What portion of the weekly budget for our family goes for food?
3. What are the various expenses a family has? What portion of a weekly paycheck goes toward each of these expenses?

CLASSROOM QUESTIONS

1. How has the price of the bread changed?
2. How much would Hamid spend for bread for each family member each month? Is that a lot? Why is Hamid upset?
3. How can you account for the difference in the price of bread in Egypt and your country?
4. Why don't we import bread from Egypt?

How should Prunella dress?

HOW SHOULD PRUNELLA DRESS?

PURPOSE

To learn not to jump at the first idea that happens.
To understand that a number in a measurement situation is not very informative unless one knows the scale of measurement.

COMMENTS

Nothing can be determined about poor Prunella unless it is known whether the thermometer uses a Fahrenheit or Celsius scale. 20 degrees F is pretty cold. 20 degrees C is quite comfortable.

The extensions give children some important benchmarks regarding temperature.

The exchange between Fahrenheit and Celsius is a ratio issue. Since the spread between freezing and boiling is 100 degrees C and 180 degrees F, there is 1 C degree for every 1.8 F degrees. That is, a rise of 10 C is the same as a rise of 18 F.

But the two scales are out of phase additively as well. The F scale is already at 32 when the C scale is at 0 (the freezing point of water). Thus to change C to F, one adds 32 and then corrects for size of degree by multiplying by 1.8. (0 C is 32 F. And 1°C is 32 plus 1.8 in Fahrenheit. 10° C is 32 plus 18 in Fahrenheit.)

Given such a procedure, challenge pupils to construct a method for going from $^{\circ}$F to $^{\circ}$C.

EXTENSIONS

What are the Fahrenheit and Celsius readings for:

1. Normal body temperature?
2. A pleasant summer day?
3. A bitter cold day?
4. A hot day in the Sahara?

CLASSROOM QUESTIONS

1. What do you think the question is really asking?
2. Do you think it's cold or warm? Why do you think this?
3. What is another temperature scale? How do they compare?
4. How do you and your classmates tell the temperature in the morning?

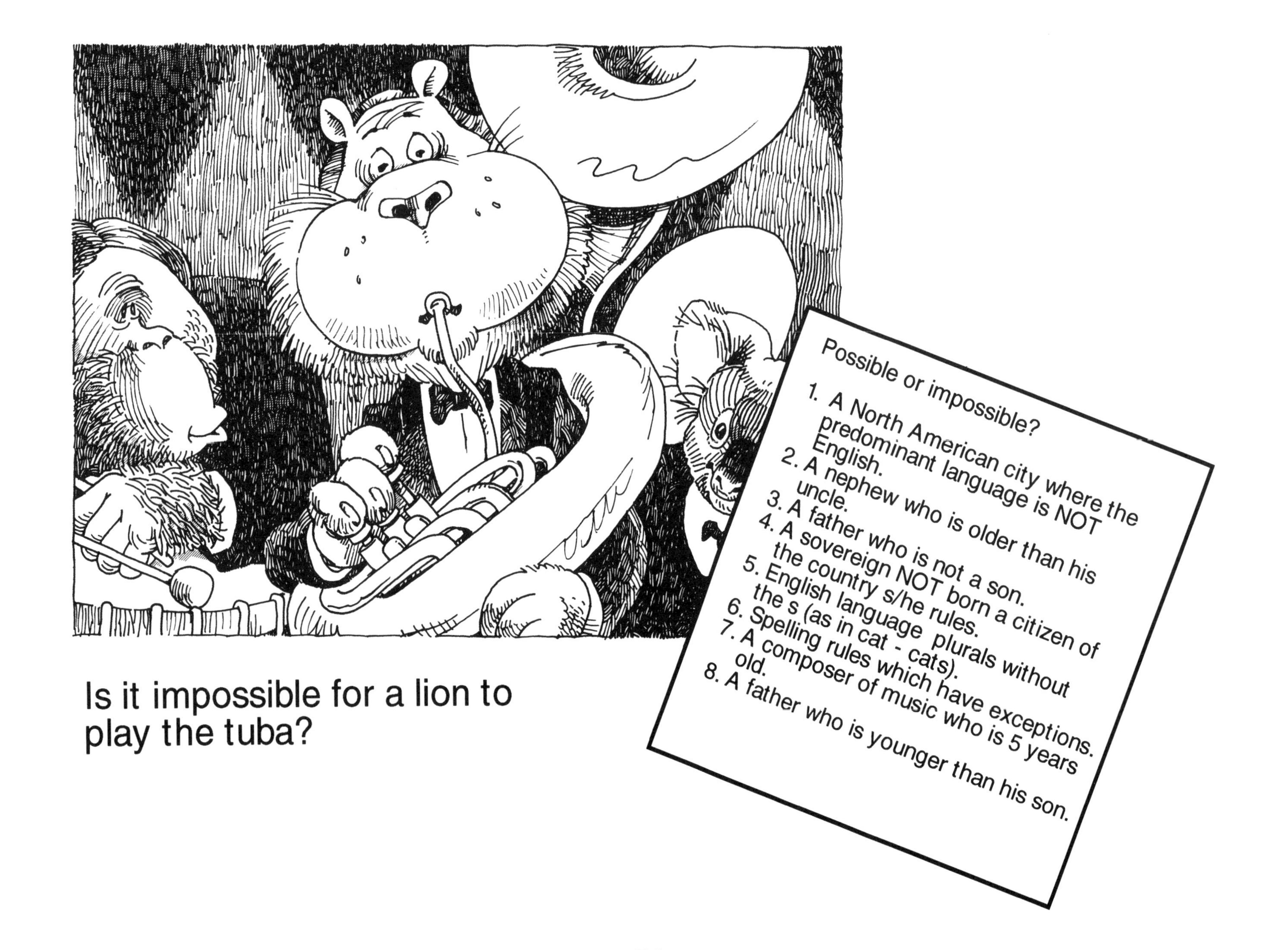

Is it impossible for a lion to play the tuba?

MISSION IMPOSSIBLE

PURPOSE

To learn about the distinction between impossible and unlikely.

COMMENTS

Question 1. Similar questions could be asked about other regions of the world.

Questions 2 and 3. What's the difference between questions 2 and 3? They both deal with family relationships.

Question 4. Did such a thing ever happen in the history of your country?

Question 5. Children. Keep going.

Question 6. "their", contrary to " i before e except........"

Question 7. Mozart. Who else?

Question 8. Younger in spirit, perhaps, but younger in age? Impossible.

EXTENSIONS

1. Encourage children to answer the question not in words but in cartoon format.
2. Encourage pupils to make up "Impossible? " problems of their own to share in class.

CLASSROOM QUESTIONS

1. How can you justify your answer?
2. Could any animal play any musical instrument?
3. How can one determine whether a situation is possible or not?

Make up a cartoon or a cartoon strip to show the consequences if there were no such thing as electricity.

WHAT IF ?

PURPOSE

To investigate the consequences of certain "what if" situations.

COMMENTS

The ability to ask, "What If" and to conjure up the consequences is a high level and important skill in real-life affairs. If one asks such questions, he or she can anticipate consequences via thought rather than be doomed to experience them in real life.

An interesting discussion of what life was like before clocks and watches were commonplace is given in Daniel Boorstin's, The Discoverers.

These questions lend themselves to pupils' inventing "What if" and "What if not" questions of their own.

An interesting book for teachers dealing with "What if" and "What if not" is Stephen I. Brown and Marion I.Walter, The Art of Problem Posing, L. Erlbaum Associates, 1983.

EXTENSIONS

Make up a cartoon or a series of cartoons to show what the consequences would be if :

1. You were 42 years old.
2. Your father was a king.
3. No clocks existed.
4. You had twice as many siblings as you do now.
5. There was no such thing as money.

CLASSROOM QUESTIONS

1. What is a consequence?
2. How is electricity important in your life?
3. How is electricity necessary to your life?
4. Ever go camping? How do you manage without electricity?
5. How was life different when people didn't have electricity in their homes? Talk to someone who lived then. Share findings.

Tell a story from these tombstones.

LIFE STORIES

PURPOSE

To generate situations consistent with given data.

COMMENTS

Given the data on Andrews and Sherman, what are the possible explanations for their differing surnames yet similar ages, when buried next to each other? What is the significance of "PRO PATRIA"?

Please note that there is no single correct story to be devised from the tombstones. Many different stories are possible but they must be consistent with the data given.

This activity was prompted by the author's discovery, in an obscure Belgian field near Namur, of gravestones erected in memory of British aviators killed in action in 1940.

EXTENSIONS

1. Draw a cartoon biography using the data from the tombstones.

2. Make a plausible biography (written or drawn) from tombstones in your town.

CLASSROOM QUESTIONS

1. Do the two people have anything in common?

2. Can this help you create a possible story for these two gravemarkers?

3. What does Pro Patria mean?

According to Forbes magazine, Charles M. Schultz (the Peanuts cartoonist) earned $3,000,000 in 1990.
How much is that per minute?

SALARY

PURPOSE

Real-life experience in computation. Experience with economics data outside children's normal context.

COMMENTS

The Schultz data are drawn from an issue of Forbes Magazine, summer 1990.

From the same issue of Forbes: William H. Cosby, Jr., the comedian and author, earned £27 million in 1990. Also the Rolling Stones were paid £22 million. Use these data for activities for children?

It might be of interest to schoolchildren to enquire about the nature of Schultz's income. Unlike most situations known to children, where a wage-earner works for an hourly or a weekly or a monthly salary, Schultz (and others in the print and electronic media) receive a large percentage of their annual income in royalties and fees for permissions and rights.

EXTENSIONS

1. The original problem calls for the computation of the number of minutes in a year. A simpler problem is "How much money per week?"
2. A slightly different problem: Suppose Schultz worked 40 hours per week for 49 weeks per year (allowing for three weeks vacation) in 1990. How much money per working day?
3. Make up some problems yourself using the Schultz information.

CLASSROOM QUESTIONS

1. How many of you read Peanuts?
2. How many people in your town read Peanuts?
3. How many papers reprint it?
4. How can one find out these data?
5. How long do you think Charles Schultz spends drawing a comic strip?
6. Try drawing a comic strip.

Some things happen regularly. The earth, for example rotates on its axis about every 24 hours.

What time regularity can you find out about the people you know?

TIME REGULARITY

PURPOSE

To obtain data and investigate real-life situations involving time regularity.

COMMENTS

It's actually false that the earth rotates on its axis every 24 hours. The figure is just shy of 24 hours. Children might want to investigate this fact and its consequences.

Once children have come up with a list of regularities as called for, ask them to explain WHY the events are regular.

The late Hungarian mathematician Tamas Varga stressed the importance of measuring the probability of various events, rather than merely treating probability as a theoretical treatment of marbles in an urn. A Varga-like question of children for each of the regularities they have cited in this investigation: What's an estimate of the probability that the event you cite will occur with regularity tomorrow? (For example, a child may cite a milkman's delivery as regular. But it's not absolutely sure that the milkman will be precisely on time tomorrow morning. What are the chances? This is a probability that can be measured.)

The teacher might want to read about (and read to the children) some of the strange situations which arose in history as the result of various calendars. The Julian calendar, for example, got so out of phase that July came in mid-winter. Take a look at Daniel Boorstin, The Discoverers.

EXTENSIONS

What time-regularity can you find about:

1. Yourself?
2. Animals?
3. The events in your home?
4. The events in your neighbourhood?

CLASSROOM QUESTIONS

1. What things do we (as a class) do with regularity?
2. How about each individual?
3. Which things that you do regularly do you need a clock to remind you about?
4. What reminds you of other things?
5. What about animals? Plants?